Managing Water Main Breaks Field Guide

AF719950

Managing Water Main Breaks

Field Guide

Kenneth C. Morgan, PE

American Water Works Association

The Authoritative Resource on Safe Water®

Managing Water Main Breaks Field Guide
Copyright © 2012 American Water Works Association
All rights reserved

All rights reserved. No part of this document may be reproduced, transmitted by any means, or stored in any database format without the prior written permission of the Publisher, except where permitted by law.

Disclaimer

The intent of this publication is to provide guidance and thoughts for consideration in the subject matter covered. It is for purchase with the understanding that neither the publisher nor the author are engaged in design, legal, system operation, or other professional services. If any of these specific services are needed, a professional agency with the required specialization should be sought out. This book does not represent an endorsement of any particular product or procedure. The editors, authors, and publisher make no claim as to the accuracy of the book's contents, or their applicability to any particular circumstance. The editors, authors, and publisher accept no liability to any person for the information or advice provided in this book or for loss or damages incurred by any person as a result of reliance on its contents.

Written permission to use the photographs in this document has been allowed as a courtesy by several water agencies across the United States. The photographs are not intended to discredit any of the agencies or to present them in a negative perspective.

AWWA Project Manager: Gay Porter De Nileon
Technical Editor: Linda Bevard
Production Editor: PerfecType, Nashville, TN
Cover Design: Cheryl Armstrong

Library of Congress Cataloging-in-Publication Data
Morgan, Kenneth C.
Managing water main breaks : field guide / Kenneth C. Morgan.
p. cm.
Includes bibliographical references and index.
ISBN 978-1-58321-860-0
1. Water-pipes—Maintenance and repair. 2. Water leakage. 3. Pipeline failures. I. Title.
TD495.M66 2012
628.1'5—dc23

2012011228

Print edition: ISBN-13: 9-781-58321-860-0
ISBN-10: 1-58321-860-2
Electronic edition ISBN: 9-781-61300-139-4

Printed in the United States of America.

American Water Works Association

American Water Works Association
6666 W. Quincy Avenue
Denver, Colorado 80235
www.awwa.org

Contents

Preface

Water main breaks are a primary cause of lost revenue and service for all water providers. Some water leaks make the evening news or the front page of the local newspaper. Others may go undetected until an unrelated excavation uncovers them or a mysterious soggy spot is identified and researched.

Main breaks occur regularly in drinking water distribution systems. Water providers experience about 240,000 water main breaks per year, according to a U.S. Environmental Protection Agency report about aging water infrastructure (USEPA 2007). Breaks interrupt service, often require costly repairs, pose the threat of contamination and the conveyance of waterborne disease, and cause customers to lose confidence in the water provider. The *Report Card for America's Infrastructure* (ASCE 2009) gives the nation's drinking water infrastructure a grade of D minus.

Some of the major contributors to water main breaks are identified in this guidebook. One of the primary issues facing water providers is the average age of the infrastructure and the importance of replacement. According to the USEPA (2002), "The useful design life of distribution system components range[s] from 65–95 years." The report further noted that the vast majority of the current underground water infrastructure in the United States was installed during the years just before through just after World War II, meaning the pipes will require replacement between the late 1990s and the 2030s.

Assessment and Renewal of Water Distribution Systems (Grigg 2004) identifies the variables that predict pipe failures. The author provides related distribution system statistics and reviews the science behind water main breaks. The Water Research Foundation is currently organizing a National Mains Failure Database to document historic information on water main breaks from U.S. utilities to be compared with data from Australia and the United Kingdom.

Another major influence on the expected life of U.S. water infrastructure is the increase in the population. The U.S. population increased by approximately 159 percent in the last half of the twentieth century, and water usage increased by 207 percent, from 14 billion gallons to 43 billion gallons. This translates to a 20 percent increase in per capita use (average use per person) nationwide. The combination of an aging infrastructure experiencing an enormous increase in its originally designed usage and the other factors presented in this guidebook directly contribute to the realities of water main breaks.

This guidebook describes the causes of leaks, how to identify them, the effective use of information documented from main breaks, and, importantly, some cost-effective ways of making the repairs or managing contractors who are performing the work.

Kenneth C. Morgan, PE
February 2012

Acknowledgments

The author would like to thank the following people for their assistance, review, and contributions to this work:

Charly Angadicheril, Assistant Water Director/Production, Fort Worth Water Department
John H. Bambei Jr., Chief of Engineering, Denver Water
Mary Beckner, Assistant Water Distribution Superintendent, City of Phoenix
Aimee Conroy, Water Services Superintendent, City of Phoenix
Neil Grigg, Professor, Colorado State University
Marcellus Jones, Distribution Systems Manager (retired), Las Vegas Valley Water District
Greg Ramon, Former Assistant Director, City of Phoenix
Marion Sanders, Training and Certification Manager, Charlotte Mecklenburg Utilities

Introduction

Numerous books, manuals, reports, and articles include discussions about aspects of water main breaks. Some go into detail, describing the science behind the breaks. Others provide effective procedures for pipe repair and identify relevant products and services. In addition, several books describe the processes of leak detection and reporting and the value of these procedures.

It is not the goal of this guidebook to reiterate the work of others or to focus exclusively on the causes of breaks or on the repair process. The focus in this guidebook is to look at these breaks in juxtaposition with their impact on the system, the customers, and others. Seeing this larger picture can aid the water provider to better manage repair, rehabilitation, and replacement options.

The information provided in this guidebook can be used for evaluating the overall impact of water main breaks and how to manage them as part of a utility's responsibility to provide a quality product to customers. Often, utilities repair their leaky pipes without ever considering the value of information they obtain during the excavation, investigation, and repair phases. Repairs are made without connecting the cause or occurrence of one leak to an adjacent leak until more damage or disruption of service pushes customers to the point of irritation and lack of confidence in the utility. These scenarios and ways to manage them are also discussed in this guidebook.

There is no one-size-fits-all solution to the repair of water main breaks, and this guidebook does not try to convey that. However the right components must be in place—the right crew size, the correct training, and the proper equipment and materials.

Each chapter concludes with a series of questions intended to help the water utility evaluate a specific issue, process, or program. These questions form guideposts that any utility desiring to provide the best service should consider.

1

Causes of Water Main Breaks

Main breaks are a reality for drinking water distribution systems. They interrupt service, often require costly repairs, pose the threat of contamination and the conveyance of waterborne disease, and cause customers to lose confidence in the water provider. Many things can cause water mains to leak or break.

- Natural disasters
- Ground and soil movement
- Defective and/or aging pipe
- Defective and/or improper fittings and connections
- Water hammer and pressure surges
- Internal or external corrosion
- Poor design or inspection of the distribution system
- Improper construction practices
- Traffic loads and insufficiently deep cover
- Unmarked or mismarked water mains

Natural Disasters

Natural disasters can disrupt water service by dislodging and/or overwhelming mains, fire hydrants, valves, and other distribution system components. Severe storms, floods, hurricanes, and tornadoes can inundate a service area so that the water utility cannot effectively provide customers with a quality product and quality service.

Some storms can destroy bridges and other structures to which distribution mains are often attached. Thus, the pipe may be pulled apart at the joints, bent, split, or even broken.

Hurricane Katrina provided an extreme example of the negative impact of weather on a distribution system. The entire New Orleans infrastructure was inundated, which challenged the utility's ability to conventionally identify the location of system components. When considering the possibility of such a severe incident, the water provider should develop plans to isolate critical portions of the system and restore them promptly as soon as conditions permit.

Weather and Environmental Changes

The effects of freezing and thawing, extreme ground movement from flash floods and storms, or the ravages of hurricanes or tornados damage distribution systems. These natural weather phenomena should be accounted for during the design and construction phases before problems arise. The maintenance, operation, and restoration of these systems should also be part of the utility's emergency response plans.

In cold climates, it is a good practice to bury the water pipe and services deep enough that the possibility of freezing is reduced or eliminated. This can provide positive benefits, such as cooler water at the tap during the summer in warm environments. During extremely hot weather, a shallow depth of cover can result in water quality problems and customer complaints.

Distribution water pipe and services buried 3 to 4 feet deep can usually aid in a utility's attempt to combat extreme temperatures. Deep installations may be more difficult to achieve in subsoil environments where rigid soil and rock exist.

If there is a substantial difference between the temperature of the water flowing through the pipe and the temperature of the surrounding soil and/or bedding material, minimal thermal expansion and contraction can occur, possibly resulting in main breaks.

Ground and Soil Movement

Another problem for underground water mains is the movement of the soil surrounding the pipe. Breaks or leaks that have washed out bedding and fill material, asphalt, or other surface treatment can cause extensive damage. This movement, which is often associated with

The era of infrastructure replacement

Like many of the roads, bridges, and other public assets on which the country relies, most of our buried drinking water infrastructure was built 50 or more years ago, in the post-World War II era of rapid demographic change and economic growth. In some older urban areas, many water mains have been in the ground for a century or longer. Given its age, it comes as no surprise that a large proportion of U.S. water infrastructure is approaching, or has already reached, the end of its useful life. . . . Prudent planning for infrastructure renewal requires credible, analysis-based estimates of where, when, and how much pipe replacement or expansion for growth is required.

Utilities [will] continue efforts to manage the number of main breaks that occur per mile of pipe rather than absorb increases in pipe failures. Utilities will strive to maintain current levels of service rather than allow increasing water service outages. Ideally, pipe replacement occurs at the end of a pipe's "useful life"; that is, the point in time when replacement or rehabilitation becomes less expensive in going forward than the costs of numerous unscheduled breaks and associated emergency repairs.

—Abridged Exceprt from *Buried No Longer, Confronting America's Infrastructure Challenge* (AWWA 2012)

flooding and storms or caused by insufficient compaction, can expose the pipe or cause it to rupture.

Defective and Aging Pipe Materials

Defective and aging pipe deteriorates more quickly than other pipe. Because of its inferior capability to convey water, compromised pipe is responsible for a greater number of water quality problems. The chemical reaction of the water, the by-products of the internal chemical reaction, and the activity of corrosive soils externally all contribute to faster deterioration of pipe materials. Once the pipe material is compromised, the pipe may be subject to multiple breaks over shorter sections or lengths. It is not uncommon for repair crews to discover multiple leaks while repairing or replacing a section of defective pipe.

Figure 1-1 identifies sections of compromised galvanized steel pipe. They contain multiple repair clamps within close proximity and

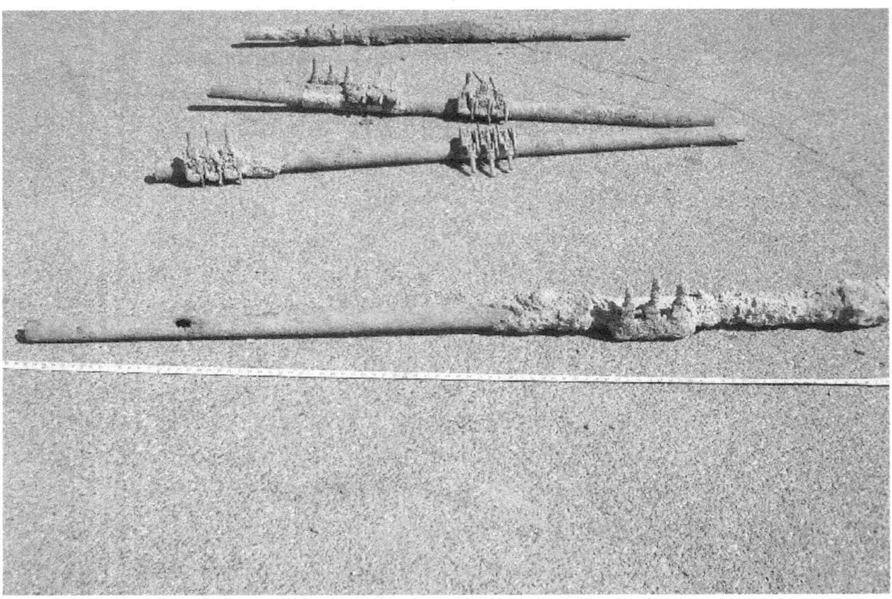

Figure courtesy of Charlotte-Mecklenburg Utilities

Figure 1-1 Sections of galvanized pipe with repair clamps

Figure courtesy of Charlotte-Mecklenburg Utilities

Figure 1-2 Corrosion buildup in galvanized pipe

yet probably were repaired at different times. These kinds of multiple repairs are time consuming, ineffective, and costly.

Tuberculation (Figure 1-2)—localized corrosion that results in knoblike mounds of rust—builds up more quickly in a compromised pipe. Subsequently, water flow through the pipe is also reduced, making it difficult or impossible to maintain adequate water pressure.

Erratic pressure surges also increase the possibility of leaks in older pipe with aging joint material. Particularly when combined with other aggravating factors, the age of the pipe can be a contributing factor to breaks and leaks. When a pipe has been in the ground beyond its design life, it is more susceptible to degradation.

Compromised pipe tends to break, burst, crack, and split. A great many of the breaks utilities experience each year can be attributed to failing pipes. Some leaks are small and will take awhile to surface. Others are evident immediately, cause major damage, and cost more to repair and mitigate. Figure 1-3 is an example of a break that occurred

Figure courtesy of Charlotte-Mecklenburg Utilities

Figure 1-3 A hole on a section of galvanized pipe

when the water pressure burst through at a point where the pipe wall thickness had been diminished by corrosion.

Defective or Improper Fittings and Connections

Leaks occur when pipe is improperly connected, bolts are inconsistently tightened, or a rubber gasket is incorrectly placed. Some leaks may be noticed immediately—for example, if a compromised fitting suddenly ruptures, blows out the pipe, and spews water. Other leaks may not be identified for a long time; pressure in a pipe may loosen the connection over time, creating a slow, hidden leak.

Water leaking from ill-fitting joints may flow along the main and come to the surface some distance away from the source of the leak. Digging at the spot where the water comes up can reveal water flowing along the pipe but no break, and so excavation must continue until the source is determined.

Water Hammer/Pressure Surges

When internal pressure exceeds the pipe's ability to maintain its integrity, the pipe blows out or bursts, often immediately. In larger-diameter pipe, this produces enormous water loss and requires a tremendous amount of cleanup. In some instances, compromised pipe material is discovered as a result of a water pressure surge in the pipe. The increased pressure typically exploits the weakness in the pipe material.

When a valve is closed suddenly or a pump fails, the flow of water is interrupted causing a series of pressure waves to flow back and forth within the pipe. These waves continue until the pressure dissipates or the waves reach a location along the main where the pipe has the ability to withstand the surge.

Water hammer or pressure surges are almost inevitable when water is transported from one location to the next. The associated problems increase when pressure surges result in main breaks. For this reason, larger-diameter pipe should be equipped with pressure-release systems to prevent catastrophic ruptures. Without these systems, or if the pressure-release systems fail, the issue is not whether a pipe will break but where.

Corrosion

Corrosion of water distribution pipe is a complex process. A report from the Centers for Disease Control and Prevention (CDC 2009) states, "When metal pipe corrosion occurs, it is a result of the electrochemical electron exchange resulting from the differential galvanic properties between metals, the ionic influences of solutions, aquatic buffering, or the solution pH."

One purpose of this guidebook is to identify some effects of corrosion on distribution infrastructure, particularly the conveyance systems. The purpose is not to explain how corrosion occurs, a topic more appropriately addressed by corrosion specialists (see sidebar on page 12).

Domino theory accounts for Boston main failure

An independent expert panel attributed the massive main break on May 1, 2010, to the failure of the studs holding the coupling on the Massachusetts Water Resources Authority's 120-in. (3,000 mm) transmission line.

The panel said that "the progressive stud-failure scenario best fits both the observed behavior of the system and the evidence retrieved from the site." The panel also considered the possibility that the failure was caused by a vertical displacement of the two pipe halves, but their review discounted that theory.

It was a classic case of dominoes: as one stud weakened and failed, the load on the other studs increased, "causing a sequential stud failure, and finally the coupling failure."

Analysis of several of the studs that held the coupling found cracks in the thread crowns and in the thread roots—likely a combination of poor manufacturing and stress during installation and startup. "Because the stresses in the studs were significantly above the allowable values, additional crack growth, environmentally assisted, occurred during operation," the report said.

After the failure, only one stud was found still in place; several fractured studs were found during excavation and repair. The relaxation of the load because of stud failures led to the loosening of the coupling and the blowout of the O-rings.

Internal Corrosion

Originally, metal water pipe was manufactured without an effective lining. Because of this, the installed pipe was highly susceptible to internal corrosion. Over time, tuberculation from electrochemical activity (the corrosive reaction of metal and water) built up inside the pipe. This weakened the integrity of pipe walls by reducing their thickness, caused tubercular buildup that decreased the inside diameter and amount of water the pipe could convey, and reduced water quality. Often a leak ensued.

Members of the panel included Zorica Pantic, president of Wentworth Institute of Technology; Ronald Ballinger, professor of nuclear science and engineering and material science and engineering at the Massachusetts Institute of Technology; and John Bambei Jr., chief of engineering at Denver Water.

The failure of MWRA's 120-in. (3,000 mm) transmission line poured 250 million gallons (950 ML) into the Charles River before the main was finally isolated and repairs could take place. The MetroWest Water Supply Tunnel, which had been put into service in 2003, replaced a 70-year-old leaky aqueduct, as part of the utility's 10-year plan to install complete redundancy for its main transmission line.

The break disrupted service to Boston and 29 other communities MWRA serves with drinking water. Within eight hours, MWRA engineers had re-routed 90 percent of the flow and found a way to make up the other 10 percent necessary to maintain pressure in the system. However, that last 10 percent required using emergency supplies and a conduit that had been out of service for decades. A boil-water alert was necessary.

The authority repaired the transmission main and restored normal flows in less than two days. By 7 a.m. on the third day, the boil-water alert was lifted.

—Mary Parmelee
AWWA *Streamlines*, 5/31/2011

Currently, cement mortar, epoxy, and other pipe linings are used to inhibit the rate of pipe degradation, in most cases successfully. However, when water gets between the lining and the pipe, the lining's protection capability is compromised. Figure 1-4 shows a thick-walled cast-iron pipe; tuberculation caused a lot of holes throughout the length of the pipe.

External Corrosion

The external corrosion of water mains typically results from the environment in which the pipe is placed, other metals it comes in contact

The experts on corrosion

The three basic forms of corrosion that commonly occur on underground ferrous metal pipelines are 1) general corrosion (uniform attack); 2) galvanic corrosion (dissimilar-metal corrosion); and 3) concentration cell corrsion. In addition to these "normal" forms of deterioration, stray current corrosion can also occur on an underground ferrous metal structure.

Do not bury copper lines next to uncoated ferrous metal lines in low-resistivity soils. The ferrous metal pipe will act as an anode and will rapidly corrode. To protect the ferrous metal line, the following items should be done in decreasing order of preference: 1) do not use copper piping, use PVC or ferrous metal; 2) coat the copper line with hot- or cold-applied coal-tar tape or polyethylene wrap and install in an insulating coupling between the copper ferrous metal line at the point of connection; and 3) coat the ferrous metal line for a distance of at least 50 pipe diameters on both sides of the connection to the copper line with cement mortoar, coal-tar enamel, or hot- and cold-applied coal-tar tape.

—Excerpted from *Corrosion Control for Buried Water Mains Pocket Field Guide*

Information on corrosion is also available from NACE International (www.nace.org), which offers presentations, publications, and training opportunities dedicated to explaining, controlling, and preventing corrosion. In addition, AWWA has two manuals of practice dedicated to understanding and controlling corrosion. Manual M27, *External Corrosion: Introduction to Chemistry and Control* (AWWA 2004), describes a list of "potentially corrosive conditions" that often result in the deterioration of water mains and their appurtenant structures. This manual also goes into detail about corrosion and how to control and prevent it in all of the affected components of the distribution system. Manual M58, *Internal Corrosion Control in Water Distribution Systems* (AWWA 2011), also has current information about rehabilitation of water mains.

with, and any other utility product in the trench. The state of matter—be it a liquid, gas, or solid—or an electrical charge can also damage the pipe's integrity.

Figure 1-5 shows an example of external corrosion attributed to the environment the pipe was placed in. Despite the thickness of the wall at the end of the pipe, its integrity was damaged because of external corrosion. Ductile-iron pipe is very durable but virtually ineffective against attack by a corrosive environment. Soil conditions directly influence the pipe material selected. Some water utilities use this information as part of an asset management program to help them determine whether a pipe gets rehabilitated or replaced. Some utilities use the information to support a requirement that coatings or polywrap be used to protect the pipe.

Pipe is sometimes placed directly on improper bedding, which can contribute to the beginning of external corrosion. Over time, the contact point of the sharp edge of a rock and the pipe could compromise

Figure courtesy of City of Phoenix Water Services Department

Figure 1-4 Tuberculated thick-walled cast-iron pipe

Figure courtesy of City of Phoenix Water Services Department

Figure 1-5 Main break caused by external corrosion

the pipe's integrity and lead to failure. The life expectancy of the pipe and other distribution components is directly related to the corrosiveness of the soil in which they are placed. Properly backfilling with a layer of select material that provides a uniform enviroment around the pipe also decreases the corrosive effects of certain soils and ground conditions. Generally, this backfill material is a mixture of gentonite, gypsum, and sodium sulfate.

Poor Design or Inspection

Distribution systems should be designed and constructed to withstand certain pressures, typically the combination of known system operating pressures and surge pressures. Under normal situations, the main should be able to operate effectively within this pressure range.

Designers of a water conveyance system must consider the operating pressure, surge pressure, possible head losses, and desired velocity. Understanding the range of pressure assumed for pipe design will help to determine the minimum diameter and thickness of the pipe. If compensation for a pressure surge is lacking, the pipe becomes vulnerable at locations where its wall may be thinner than the minimum required. If the pressure exceeds the design rating, a main break will result.

Also critical to the design of large-diameter pipe is the proper distribution of air-release valves. These fittings, placed at the high points along the main, provide a place for air in the line to be removed without damaging the pipe.

Improper Construction Practices

An improperly supported pipe moves during a pressure surge. Over time, this movement results in the separation of the pipe joint and, ultimately, a leak. Depending on when it happens, this type of leak can delay the completion of other construction; the area must be dried and additional material brought in to fill the hole.

If bolts are not securely tightened on fittings, leaks result. The internal pressure of the water exacerbates the looseness of the bolt, eventually compromising the connection and allowing water to escape.

On most gasketed push-on joints, an improperly placed rubber gasket will not seat effectively, and the result is a leaking joint. Such a leak may not be discovered until the pipe is hydrostatically tested, although it may eventually become evident without such testing. Improperly seated gaskets can often take as much time to repair as a pipe blowout, depending on location, depth of cover, the presence of other utilities, and adjacent traffic conditions.

The placement of bends, tees, and similar fittings that change the direction of flow should be accompanied with thrust blocks or restraint fittings to resist vertical thrust. If such support is not provided, any major surge in pressure experienced by the connection may cause movement in the pipe that separates the fitting from the main.

It is particularly critical that the joints be welded and tested properly during the installation of welded steel pipe. If this is not done, the pipe may not pass the required pressure test, eventually resulting in a leak. In larger-diameter pipe, this can be catastrophic.

Inadequate compaction can also cause water main breaks. If fill material is placed without regard to its importance or consequences, the result is often lack of adequate support around the pipe. This happens more often than is realized; breaks associated with inadequate compaction are usually attributed to something else, such as ground movement or water hammer.

Because water flows through pipe primarily by pressure, distribution systems often experience surges in the lines. If the distribution system is not appropriately designed and constructed, joints can separate because of surges of high-pressure water, resulting in a water main break.

Another important construction consideration is the soil support surrounding the pipe, particularly while the pipe is under pressure. If the soil is not compacted closely against the pipe, gaps of air will occur, and the pipe walls will lack support that helps protect against pressure surges. This can result in a break or a separated joint.

Traffic Loads and Insufficiently Deep Cover

Insufficient depth of cover above the mains can contribute directly to the breaks and leaks when there is a heavy traffic load. Commercial vehicles traveling over distribution pipe can create problems at connections and fittings. The pipe should be buried deeply enough and have a sufficient factor of safety to counteract traffic-related stresses of weight and movement.

Traffic loads are a greater problem on bridges and other structures to which water distribution pipes are connected. The traffic loads generate pipe movement (usually at connections and fittings) that causes mains to crack and ultimately break. The force of the vehicular loads can displace gaskets and loosen bolts, fittings, and the pipe itself.

Once gaskets or fittings are displaced, slow leakage can occur at that site, and rust may form on the fitting and adjacent pipe. The continuing movement of the pipe from traffic load can further aggravate the leak. Also, when pipe supports on bridges and associated structures are improperly spaced and secured and become compromised or wear out, the result is usually a main break.

Unmarked or Mismarked Water Mains

The water utility is responsible for providing the location of the underground infrastructure at the request of an excavating party. If it is an emergency request, the utility should dispatch someone to locate the position and mark the road surface as a reference point. In several states, such marks are considered valid within ± 2 feet. If the utility does not have an underground line in the dig area, the locator will often mark that the area as "all clear."

If the excavator receives an all-clear from the water utility, the assumption is that no distribution system infrastructure is present in the dig area and the excavator can proceed quickly and without particular caution. If the excavator encounters an unidentified water main, the excavation equipment could slice through it or dig it up, likely breaking or damaging the pipe in the process.

Such an incident usually slows down the excavator's schedule and can add time to the project. Water may enter the excavation, and time must be taken to repair the main and mitigate any related damage. The main break shown in Figure 1-6 occured because a locator

Photo by Kenneth C. Morgan

Figure 1-6 Main break that damaged other utilities

improperly marked the main. The water utility subsequently bore the costs of the repair.

Construction Damage

In some instances, the contractor may cause the break or leak. This may be the result of an inattentive operator or a mismarked water main or service. In some cases, a handful of less reputable contractors may not notify the water provider of the breach until they perform their work or they may not inform the utility at all. An inspector with knowledge of the water system should be dispatched to the construction site during any excavation to monitor the contractors' activities. Timely response to these breaks is critical and should be effectively managed to minimize the service interruption.

Figure 1-7 is an example of a construction site that has been damaged by a leak. Repair and site restoration delayed completion of the project. Notice the section of pipe displaced by the leak.

Photo by Kenneth C. Morgan

Figure 1-7. Leak that negatively impacted a construction project

Conversely, activity by other utilities can damage water mains. During construction of buildings, bridges, or roads, a water main can be struck or punctured, resulting in a break. When this occurs, completion of the project is delayed. Not only must the pipe be repaired, but often the affected area must be dried out, or the oversaturated soil must be removed and replaced. Depending on the size of the break, this can take considerable time, resulting in cost overruns and additional expenses.

QUESTIONS TO CONSIDER

1. Does the utility document the causes of main breaks?
2. What are the percentages and numbers of the various causes of breaks in the system?
3. Does the utility have programs for monitoring, evaluating, and reducing leaks?
4. On average, how many main breaks does the utility have per day? Per week?
5. Is the number of unmarked or mismarked water mains and services monitored by the utility?
6. What is the utility's approach when the break/leak is caused by a contractor or another utility?
7. Has the utility collected baseline hydraulic and water quality information for critical areas of the infrastructure?

2

Impact of a Water Main Break

When a main breaks, the utility must mitigate the damage to the water system, the environment surrounding the break, the quality of the water, and customers' concerns. The main may be adjacent to other underground utility services, such as gas or electric, causing the repair or replacement to be more difficult and time consuming. In some instances, a representative from the other utility must be present during the repair, or the relocation of another utility line may be required before or shortly after the repair or replacement of a water main.

Some water mains are located in easements within customers' property, and when restorative work is performed, it can affect the condition of the driveway, sidewalk, or yard. Mains located under medians, trees, or decorative surfaces will increase the time and cost of repairs because of the additional work associated with mitigating these conditions.

If a water quality concern arises out of the repair of a section of pipe, it could take a while to resolve the issue. If required, the results of samples take time, and some of the necessary responses take even longer and often involve communicating with the media, the public, and officials from regulatory agencies. One of the biggest challenges is eliminating the possibility of cross-contamination as a result of the break or repair process.

Impact on the Provider

No matter how well trained the staff may be or how much equipment a utility has, consumers usually judge the water provider by its response to a crisis or emergency. For this reason, preparedness is of the utmost importance. Any time a distribution system experiences many main breaks over a short period of time, a large-diameter pipe breaks, or any pipe breaks at a critical location, the second-guessing and scenario-pondering begin to increase. A utility that is more reactive than proactive may not have anticipated the event, and it could thus be hindered from providing an effective response. A proactive utility is more likely to respond effectively because it places importance on component or system monitoring, leak audits and surveys, properly managed operation and maintenance programs, an effectively trained staff, and on having enough repair parts and replacement materials on hand to respond in a timely manner. This preparedness helps to

Main break sends people to shelters

DUNDALK, MD. Baltimore County Public Works officials reported a 72-in. (1,800-mm) water main that broke in the community south of Baltimore, flooded three other communities, destroyed a large section of a major highway, and left more than 840 homes without power. A shelter was opened for residents, and the Red Cross distributed 2,000 meals. Photos of the event were shown to the Senate the following week during debate on appropriations for water infrastructure as proof of the need for increased water infrastructure funding. The Senate appropriated $1.4 billion for the Drinking Water State Revolving Fund.

—*AWWA Streamlines* 09/29/2009

reduce the negative impact of a main break to the utility and, ultimately, the customer.

Every water utility should be prepared to meet the challenge of any crisis, whether human-caused or natural, that reduces its ability to provide water to the customers. One of the best ways to be ready is with a well-thought-out plan for most scenarios. AWWA Manual M19, *Emergency Planning for Water Utilities*, provides vital information to assist water providers in developing these plans.

It is virtually impossible to have a predetermined response for every situation. Nevertheless, consistent portions of the response plan should be established no matter what the crisis. These responses should be rehearsed and understood by the appropriate water utility employees.

Impact on the System and Staff

A break in a water line can affect the distribution system in various ways. Operating pressure could drop below the minimum required by regulation, possibly resulting in the introduction of contaminants into the distribution system by the vacuum effect or by backflow of water from the adjacent area into the damaged main or service. If a booster pump, pressure-reducing valve, or other water conveyance component downstream of the break is affected, the disruption of service will require the component to be monitored and adjusted, and possibly repaired. If multiple breaks occur over short distances, or a major blow-out occurs, system modifications may be necessary that impact more of the infrastructure, and water may need to be rerouted to provide service to affected customers.

When a system component fails without warning, water distribution operators must immediately focus on restoring service. Most utilities have staff dedicated to handle a typical break. They may be able to restore service within a short period of time, or it may take several hours. In some cases it takes days, because of the impact of the outage, the repair, and the restoration processes. It may be necessary to mobilize additional crews, alter work schedules, or employ a contractor to respond to more severe breaks.

Figure 2-1 shows a water main break that affected the water system, its local customers, and the community. The break not only disrupted the flow of water and traffic, it also undermined consumer confidence. When damage is this extensive, the utility may need to provide bottled water to ease customers' frustration while repairs are made.

The water provider should document the amount of water lost from the break and ensure that there will be enough supply to replace it. This becomes a significant concern if the lost amount is enormous. If the lost amount depletes the stored supplies, the provider may need to rely on an adjacent water provider or a bottling utility for an alternate supply.

Many water systems have equipment that automatically tracks the amount of water treated, pumped, stored, and used so that a balance can be determined. This balance helps to reflect the water provider's effectiveness in managing its system.

If a water quality concern arises out of the repair of a section of pipe, it could take awhile to resolve the issue. Samples may be required, and laboratory testing takes time. If the samples come back positive for a contaminant, regulatory actions, such as public notification, are initiated, and usually involve communicating with the media, the public,

Water pressure and leakage

Operating pressures not only have a major effect on the amount of water escaping from active leaks but also a surprisingly large influence on the rate of generation of new leaks. Water main break and service leak frequency increases rapidly when high pressure is encountered, either as pressure surges or when operated at continuous high pressure. Operating the distribution system at a steady level of pressure sufficient to sustain the desired level of service to customers, but not at excessive levels, can garner savings from leakage reduction and results in less stress on distribution system infrastructure.

—Excerpt from M36, *Water Audits and Loss Control Programs* (AWWA 2009)

Photo courtesy of City of Phoenix Water Services Department

Figure 2-1 The impact of a main break on the immediate environment

and regulatory officials. One of the biggest challenges is eliminating cross-contamination caused by the break or the repair process.

Impact on the Customer

Few tasks are more difficult for water utility personnel than issuing a boil-water notice or telling customers about a temporary loss of pressure or service. The primary goal of any water provider is to supply a quality product at an adequate pressure at all times to those in their service area. When problems occur, customers expect them to be handled immediately, and their confidence is diminished when a main break requires time-consuming repairs. It is important to keep customers informed of the progress of the repairs and their impact on service.

Isolation versus immediate repair

It is sometimes beneficial to isolate the area of the main break for repair at a future time. This should be considered if some of the following criteria are met:

- There are no customers on the section where the break is isolated.
- Isolation does not reduce flow or pressure to customers adjacent to the area of isolation.
- The repair will be scheduled, and service restored, in a timely manner.
- Some or all of the components necessary for repair are not on hand.

Isolating the main break for future repair has drawbacks if an extensive amount of time elapses before the break is isolated and if the disruption of service to the adjacent customers causes more than the normal difficulties.

Critical to the repair effort are the communication and coordination required so that the customers are not further inconvenienced. If the break is on a critical large-diameter pipe, water treatment plant staff should be notified as well. Engineering staff can often assist by providing design and as-built information. If fire hydrants will be out of service for any length of time, the fire department should be notified of their locations and the approximate times of the outage.

Some key actions to maintaining customers' confidence are to quickly and responsibly attend to their water concerns, investigate any service disruptions in a timely manner, provide an alternative water source when necessary, and restore service in a reasonable time without having to interrupt the customers again within a short period beyond the repair.

There is a belief that as the opinion of a few select customers goes, so goes the opinion of the community. If a break receives media coverage, people in unaffected areas often want to know if it will affect them. The larger the event and the more the system is affected, the more customers will have concerns. The communication strategies necessary for the different levels of repair are presented in Chapter 5.

Another way water providers maintain their customers' confidence is to communicate with them in real time when changes affect their distribution system. They should be notified of more than main breaks and should also be informed about improvements to treatment and distribution processes. This practice makes customers aware of the utility's desire to be current with the best management and operational practices throughout the industry.

Impact on the Community

Main breaks impact communities in other ways as well. Some breaks can interfere with traffic flow, adding time to the workday commute. The break shown in Figure 2-2 had such an impact. Fortunately, no one was injured when the main ruptured or the crater subsequently appeared.

Photo courtesy of Denver Water Department

Figure 2-2 A main break that impacted local traffic

The economic cost of leaks and breaks

Real [water] losses include water that has been extracted from a water resource source, treated, energized, and transported a distance before being lost from the distribution system. Because these quantities of loss occur in addition to the water successfully supplied to customers, real losses effectively impose on the water utility excess extraction, treatment, and delivery charges, and/or excess imported water purchase charges. Treatment and delivery costs include the variable costs to produce the water, or the costs to produce the next million gallons (or other standard increment) of water. If the water supplied is purchased from a neighboring water utility, the purchase unit cost should be applied.

Other long-term costs also exist for real losses. The cost of wear and tear on treatment and pumping equipment might be taken into account in the supply costs, particularly if real losses are high. Additionally, because real losses represent volumes of water taken from a source that do not generate a benefit, these losses could also be assessed costs relating to their environmental, economic, and social impacts.

Because leakage volumes [and loss from main breaks] are a considerable portion of system input for many water utilities, expansion of water supply infrastructure might be deferred if successful leakage control is achieved. Likewise, better use of existing resoruces may ease drought restrictions or allow economic development to occr without exploiting new water resources. Reducing leakage volumes [and main breaks] results in a corresponding reduction in the operation of equipment, thereby extending the interval between scheduled maintenance.

—Abridged excerpt from M36, *Water Audits and Loss Control Programs* (AWWA 2009)

Businesses and schools are often negatively affected when water main breaks disrupt their operations. If a repair cannot be performed in a reasonable amount of time, the employees of the business or the students and staff of the school are usually released. This inconveniences the customers of the businesses or the family members of the students. If possible, the needed repair should be delayed until after business hours, after school is dismissed for the day, or the weekend.

Knowing the location of critical customers assists a water provider when determining the level of service disruption necessary for the repair of a main break.

Impact on Other Utilities and Agencies

Most municipalities have utility corridors that several utilities share. Infrastructure for more than one utility is often in a close proximity in these corridors or rights-of-way. This close proximity creates challenges when water mains break or leak. The damage to other utilities is often compounded based on their physical closeness in the affected area. The water pressure from a main break can often disrupt other utilities' services. This disruption can be minimal or it can destroy another utility's equipment, as shown in Figure 2-3.

Unmarked or mismarked utility lines are often the cause of costly and time-consuming problems. To reduce their frequency and number, the utility must maintain up-to-date maps and as-built drawings and modify them whenever a main is extended, improved, relocated, or repaired.

QUESTIONS TO CONSIDER

1. Has the utility recently experienced any major disruptions to the service you provide because of a main break?
2. What arrangements, if any, does the utility have with emergency responders in notifying the utility of main breaks?
3. Does the utility have requirements or guidelines for requesting water main locations?

Photo courtesy of City of Phoenix Water Services Department

Figure 2-3 A main break that disrupted the service of a communications utility

4. Identify the impacts of main breaks on the system and the customers.
5. How does the utility communicate critical information regarding leaks to the customers?
6. Is the utility's customer notification process effective?
7. What is the utility's plan for providing an alternative water supply during the repair process?
8. What is the utility's process for handling claims?

3

Identifying Water Main Breaks

Information about a leaking or broken water main comes from various sources, including utility employees, customers, leak detection programs, contractors, and emergency responders.

Utility Employees

As they perform routine tasks in the field, water utility employees frequently find evidence of main breaks or leaks. These workers regularly operate and maintain system components in their service areas, so they recognize when something unusual has occurred. In this way, many minor breaks and leaks are identified before they can cause more severe damage or increase the length of a service interruption.

Customers

Customers are often helpful in noticing leaks. Usually they observe a wet spot in the road or water bubbling up through a grassy area, water flowing on the surface or spraying in the air, or a small hole (or even a huge depression) in the pavement. Their observations are sometimes accompanied by the loss of water pressure or water service, which may be the initial reason a customer calls the water utility.

A large volume of calls usually means that customers from a broad area have been affected, unless the calls are from a high-rise building or apartment complex where customers are heavily concentrated. The water provider's response should be timely and effective. The utility should respond immediately to customers' water needs and inform them of the overall repair plan. This helps build customer confidence and may provide support during rate increase meetings.

Leak Detection/Audit Programs

Water utilities that use leak audits, surveys, and detection programs proactively identify breaks, reducing the amount of water loss from

Controlling the leak

The first consideration in dealing with the leak should be to get the water loss under control by complete or partial shutdown of the line. Some repairs can be accomplished while the pipeline is still under pressure. Maintaining a positive pressure in the line will ensure that backflow from cross-connections will not occur. Where damage to property is occurring or is likely to occur, water must usually be completely shut off.

If there is no immediate danger to life or property, all valves on that line except one should be closed to reduce the flow. Customers who will be without water should then be notified of approximately when the water will be shut off and advised to store some drinking water if they will need it during the outage. Special attention should be given to buildings with sprinkler systems, stores with water-cooled refrigeration units, industrial water users, and large users such as hospitals. These customers will have to take special steps to prepare for the water being off.

—Abridged excerpt from *Water Transmission and Distribution* (AWWA 2010)

System leak detection program

There are three methods to find leaks:

Audible leak detection. This method uses electronic listening equipment for detection. Pressurized water that is forced out through a leak loses energy to the pipe wall and to the surrounding soil areas. This energy creates sound waves in the audible range, which can be sensed and amplified by electronic transducers or, in some cases, by simple mechanical devices. The sound waves are then evaluated by an individual trained in leak detection who can determine the exact location of the leak, based on the pitch or frequency. This specialist conducts an initial listening survey of the entire distribution system and records all suspect sounds. If these sounds are heard again when rechecked later, leaks are pinpointed.

Zone flow measurements. This method can reasonably be undertaken as an extension of the water audit, or, in some cases, directly as a leak detection method. Its purpose is to determine whether or not a sector or zone of a water system is suffering major leakage. To effectively conduct a zone flow measurement, a utility must maintain good maps, have valves located at zone control points, and provide a tap in the main for the recording pitometer.

Normal course of operation. This method involves discovering leaks accidentally during the normal course of operations and maintenance, for example, in a valve-exercising program. Meter readers have the opportunity to check for visible meter box leaks when reading the meters.

—Excerpt from *Utility Management for Water and Wastewater Operators* (Bloetscher 2011)

their systems. These programs are routinely scheduled and performed as part of a utility's operation and maintenance programs by utility staff or by a contractor. One goal is to locate leaks and then schedule and perform repairs before leaks become catastrophic, particularly in critical areas or in locations with a large volume of customers. This also greatly reduces the cost of the repair or replacement. Another goal of these programs is to minimize the exposure of the system to the possibility of contamination.

Without an audit program some leaks may never be discovered, and water quality and quantity may be placed in jeopardy. Leak audits and detection programs also assist a utility by accurately identifying a portion of their water losses. Whether the nonrevenue water recovered as a result of these efforts is minimal or extensive, the audit program truly reflects the desire of a water utility to properly manage its product.

Generally, leak detection methods are used when a leak has been difficult to locate through routine observation, or the leak is in an area where it would be costly to just open up the ground to look for it.

Photo courtesy of City of Phoenix Water Services Department

Figure 3-1 Break in a 36-inch main discovered by leak detection, avoiding a blow out

Characterizing leakage events

All drinking water utilities encounter reported leaks. Utilities whose only leakage control activity is to respond to reported leaks are operating a reactive leakage management program. Systems that additionally seek to identify unreported leaks are operating a proactive leakage management program. For many utitlities, most leakage losses over the course of a year occur from unreported leaks. For those systems with a reactive leakage control policy, it is likely that they are controlling only a minority of the leakge occurrences in their distribution systems.

—Excerpt from M36, *Water Audits and Leak Detection* (AWWA 2009)

Accurately performed, leak detection methods such as acoustic listening devices and in-line sensors have proved repeatedly that pinpointing a leak costs considerably less than disrupting a large surface area to look for it. Figure 3-1 is an example of a leak on a section of concrete pipe that was discovered by the detection process and was repaired before it became a catastrophic break.

AWWA Manual M36, *Water Audits and Loss Control Programs*, covers various leak detection methods and their value to water providers.

Contractors

Sometimes contractors who are working for the water utility or another utility discover main breaks or leaks. Water may be flowing into the contractor's excavation site, or the contractor may have been digging alongside the water pipe and noticed water. Most of these contractors immediately notify the water provider, which dispatches a repair crew. If the contractor is allowed to make the repair, a representative of the water utility should be on site to supervise the work.

Photo courtesy of City of Phoenix Water Services Department

Figure 3-2 A high-pressure main break attended to by first responders

Emergency Responders

Figure 3-2 shows the result of first responders identifying a leak and notifying the water utility so that an effective traffic plan was put in place to prepare for the repair. This effort kept the site safe and reduced the chance of a vehicle getting damaged or pedestrian getting harmed.

Emergency responders and water personnel can practice procedures in tabletop exercises or field demonstrations so that both parties understand their roles in the mitigation of various water-leak scenarios.

First responders and water distribution personnel cooperated in the situation shown in Figure 3-3, when a pickup truck sank into a hole created by a main break. This break could have caused serious injury to the driver and passengers. An alert police officer using common sense assessed the immediate danger to pedestrians and vehicles; it was easier to allow the vehicle to be damaged than to risk harm to the passengers as well.

QUESTIONS TO CONSIDER

1. Does the utility have an effective water audit and leak detection program?
2. Does the utility have an effective pipe evaluation program?
3. Does the utility have an effective mechanism for identifying main breaks?
4. What does the utility consider to be critical information for the leak report form, and how is the information captured?
5. Does the utility have accurate existing infrastructure information?

Photo courtesy of City of Phoenix Water Services Department

Figure 3-3 A main break that claimed a vehicle

6. How is the leak information database accessed?
7. How much of the defective infrastructure is replaced each year?
8. Does the utility have an arrangement with first responders for dealing with breaks that disrupt traffic and commerce?

4

Repair Site Safety

Safe repair of a water main break is of the utmost importance to the repair crew, the utility, and ultimately the community. Several critical safety factors must be considered when making repairs. The repair process itself, including prioritization, evaluation, and other related topics, will be covered in chapter 5.

Emergency Responders

When a major leak disrupts traffic and interferes with local business operations and the normal flow of water to customers, emergency responders—usually the police or fire department—are often the first to arrive. These trained professionals are equipped to isolate intersections, stop traffic flow, and clear streets in residential and business areas.

If a large-diameter line has ruptured, the force of escaping water can cause great damage and threaten unsuspecting pedestrians or drivers. These types of breaks have the potential to make water utilities liable for huge sums of money in repair costs and claims. On a major thoroughfare, quick reaction by first responders can avert the possibility of damage to vehicles and injury to passengers. Figure 4-1 shows a main break that emergency responders effectively isolated until utility personnel could mobilize and begin the restoration process.

Most municipalities require a plan for traffic control when repairs are made in or adjacent to a roadway. The placement of traffic barriers, cones, barricades, and so on provides a controlled environment in which the crew can proceed with operating valves to isolate the main, making repairs, or resurfacing streets. Flaggers are sometimes employed to expedite traffic flow around the work zone while allowing the work to continue safely. If the street must be entirely closed for safety reasons, drivers would need to detour.

It is critical to set up the traffic control plan prior to beginning the excavation. During the course of the work, there will likely be a

Photo courtesy of City of Phoenix Water Services Department

Figure 4-1 A major main break that impacted traffic

backhoe, a dump truck, a utility vehicle, possibly other equipment, repair crews, and customers moving around vying for the same space at the same time.

Repairs by Water Utility Staff

The water utility should be prepared to respond to any incident, routine or emergency. Training is a vital part of this readiness. Staff members should be trained using standardized procedures that are regularly reviewed, taught, and practiced so that everyone is familiar with the tasks to be performed. Training should specify consistent ways to perform all necessary repair functions. If crews have been established for particular tasks—valve operation, service line repairs, main repair and replacement, and so on—consider cross-training employees to balance knowledge and expertise throughout the organization.

Broken or leaking water mains are repaired either by water utility employees or by contractors. Water utilities that provide their own repair crews usually have procedures in place to expedite the repairs. The size of the repair crew, the roles of each person on the crew, the fittings and materials used, and so on, are vital parts of restoring water distribution. Whether the crew is from the water utility or is contracted, workers should use all job-site and trench safety procedures when responding to breaks and leaks.

Water mains do not always break during the 8-hour day shift, so crews should be prepared day or night. Utilities that operate on a 24-hour basis may have repair crews at work no matter what time a repair call comes in. Some water providers use an emergency standby schedule to guarantee timely response by an established repair crew. For this to be effective, the utility must maintain an updated list with contact information for crew members.

Repairs made by utility staff should be documented and the process evaluated continually. This provides the water utility with a mechanism to confirm the effectiveness of the repair staff or provides valuable information to improve future main break responses.

Crew sizes required for various tasks and the functions of each member should be documented so that all repair teams accomplish the same tasks the same way every time. These standard operating procedures increase efficiency.

Repairs by Contractors

Water utilities unequipped to make repairs using their own workforce usually contract out this effort. Service agreements with contractors often cover the types of repairs to be made, expected time lines for completion, customer notification procedures, provision of alternative water supplies, and other pertinent topics necessary to successfully complete the repair and maintain water service.

The contractor performing repairs to the water distribution infrastructure should be monitored by the water utility so that work is done according to the construction and materials requirements of the water provider or contract. The lengths, sizes, and locations of all repair components should be documented as well to ensure a complete record of the repair.

The repair agreement should guarantee all aspects of the work. It should include language that identifies the evaluation of and recommendation for the pipe being repaired, the soil conditions around it, and the particulars of the surface restoration. If the utility uses a blanket contract with the same vendor for multiple repairs over time, the utility should specify enough general information about leak repair in the agreement to allow for flexibility but maintain provisions that guarantee the completeness of the specific repair with an addendum that notes the particulars of that repair.

Contract work should benefit the agency performing the work as well as the utility for which the work it is done. Work progress must be monitored by a responsible party familiar with the specifications of the repair agreement. If contractors are improperly or inadequately monitored, another break could result or customers could be without service longer than expected. The contractor should provide a report that includes a schedule, the materials used, the number of persons performing the job, the equipment used, and any special conditions.

Safety Procedures

The practice of a safe response to the repair of a water main break is of the utmost importance to those performing the work, the utility, and ultimately the community in which they serve. Most utilities are required to have a competent person at each job site. This individual is responsible for monitoring the condition of the trench, maintaining the overall cleanliness and safety of the job site, and directing the correction of any possible hazards or infractions. Even though safety at the worksite is everyone's responsibility, it is helpful to have this added oversight.

Several actions should be standard procedures, as they are critical in maintaining a safe work environment. These key actions are described here.

Barricades

Setting up barricades and safety cones provides a secure barrier for the workers to move about when operating valves for isolating the main, making repairs, or resurfacing the impacted streets. In some cases, flaggers are employed to allow the work to continue safely

Traffic control safety

Barricades, traffic cones, warning signs, and flashing lights are used to inform workers and the public of when and where work is going on. These devices should be placed far enough ahead of the work so the public has ample opportunity to determine what must be done to avoid obstructions. If necessary, a flagger should be used to slow traffic or direct it. Everyone involved in the work should wear a bright reflective vest.

Approved traffic safety control devices that should be used for obstructing roadways are described in detail in *Manual on Uniform Traffic Control Devices for Streets and Highways*, prepared by the US Department of Transportation, Federal Highway Administration.

Most states have also preapred simplified booklets describing work area protection that should be used during street and utility repairs. . . . Water utility operators should be aware that they could be held liable for damages if an accident occurs as a result of utility operations that were not guarded in conformance with state or federal procedures.

—Excerpt from *Water Transmission and Distribution* (AWWA 2010)

Photo: Bryan Bechtold, AWWA

Figure 4-2 Traffic moves past pipe repair work in a coned area

while expediting the flow of traffic around the work zone. An important consideration for the safety set up is to protect the work crews as well as provide them with an effective process to get equipment in and out of the work area. Figure 4-2 shows a coned work area.

Water and Air Pressure

A vital part of the repair process is to reduce or relieve the pressure on the water main that requires repair to eliminate the possibility of additional damage to the surrounding area or even harm to one of the workers. This process affects the customers connected to that main.

Many water providers use a log-out/tag-out procedure for valves when they respond to main breaks. In this process, the persons responsible for isolating the area to be repaired close and open valves as necessary and leaving an established notice in the valve box with their contact information to make sure they are consulted before the valve is opened or closed. This prevents the possibility of additional damage or injury. It also helps the repair crew keep track of when certain valves have been operated.

It is sometimes beneficial to maintain reduced pressure on the line until the pipe has been excavated and the leak has been pinpointed. Certain breaks or leaks may be on the bottom of the pipe and are difficult to identify unless positive pressure ensures continuous flow. If this process causes safety concerns, a further reduction in pressure should be considered. In the case of larger-diameter pipe and in the interest of safety, the pressure should be completely turned off during the repair.

Properly maintained fire hydrants can be used to help relieve the pressure on the main, to check how effectively the affected main has been isolated, and to remove air from the line while recharging it. Water flow from the hydrant can also be used to collect and confirm hydraulic and water quality data. Flow and pressure measurements taken after a main break may vary from previously documented baseline values. Lack of a chlorine residual in the immediate area may indicate that contaminants have been introduced.

Air can be released from the line under repair through service lines adjacent to a main break. These service lines can also be used to move chlorine through that portion of the system if no fire hydrants are strategically located to provide this level of support. It is imperative that the service line used for air release belongs to the water provider. Otherwise, the water utility may have to replace fixtures and/or appliances at a residence or business as a result of damage by waterborne debris or trapped air.

Shoring

Proper shoring should be in place during excavation and before workers enter the area of the break. This will be the case particularly when other utilities are adjacent to the pipe or if there are other restrictions—such as traffic restraints, curbing, or other structures. Figure 4-3 shows a trench box being moved into place to shore up an excavation.

Once the shoring has been secured in place, workers can enter a safe work zone that is free of the possibility that the surrounding soil will collapse inward. The use of shoring for excavations deeper than five feet is an OSHA requirement. An exception for a deep excavation is to slope the sides of the hole to reduce the severity of the confined space and allow for easy egress (Figure 4-4). When the excavation is large and the sides are sloped, effective egress is possible.

Photo: Bryan Bechtold, AWWA

Figure 4-3 A trench box provides solid walls for a narrow excavation

Repair Equipment

The proper equipment, or lack thereof, to maintain and repair a distribution infrastructure affects the amount of time required to restore water service. Water providers should determine the critical components and equipment required to maintain the level of service they have committed to provide. Utilities should have an ample supply of repair components. If the utility does not have something in stock, a previously established interagency agreement may allow the water provider to quickly acquire what it needs from nearby utilities.

Another important consideration is the amount of time it would take to have the component shipped in if necessary. An unreasonable delay not only holds up repairs, it also affects customer service and perception. The water provider should consider expanding the supply of parts it keeps in stock. For larger, more expensive components, the supply should be replenished as items are put in the field.

A list of equipment suppliers or contractors should be maintained so access to items the utility does not own can be acquired in a timely manner. In addition, the utility should determine the most effective excavating equipment necessary to make the repair. The lifting capacity, the reach, and the bucket size, among other things, all influence the selection of machinery necessary to make an appropriate response.

Equipment should be chosen based on the following criteria:

- The depth of the main(s)
- The various sizes of the mains
- The different ways in which the equipment is used
- The size and number of crews making repairs
- The budget the utility has for equipment purchase, maintenance, and so on

Photo courtesy of Denver Water Department

Figure 4-4 A huge excavation with sloped walls

Energy isolation

The established procedures for the application of energy control (the lockout or tagout procedures) need to cover the following elements and actions and shall be done in the following sequence:

1. Before shutting down or turning off a machine, equipment, or other source of hazardous energy, such as high-pressure hoses, the authorized or affected employee who will perform this task must have knowledge of the type and magnitude of the energy, the hazards of the energy to be controlled, and the method or means to control the energy.
2. Machine or equipment shutdown must follow the procedures established for the machine or equipment. An orderly shutdown must be used to avoid any additional or increased hazards to employees as a result of the equipment stoppage. In the case of high-pressure sources, block the fluid (gas or liquid) flow in hydraulic or pneumatic systems. Vent fluids from pressure vessels, tanks, or accumulators, but never vent toxic, flammable, or explosive substances directly into the atmosphere.
3. Machine or equipment must be separated far away from the energy-isolating devises that are needed to

Personal Protective Equipment

Hazards to workers in the excavation must be evaluated. When soils are oversaturated, it is difficult to maneuver in certain soils to dig out an area under the pipe, carry fittings, or make repair connections. Workers often injure their knees or lower backs when they try to move too quickly.

control the energy to the machine or equipment. That is, they shall be physically located and operated so that the machine or equipment is isolated from the energy sources.

4. Lockout or tagout devices are to be fixed to each energy-isolating device by authorized employees. Lockout devices shall be fixed in a manner that will hold the energy-isolating devices in a "safe" or "off" position. Tagout devices must be affixed in such a manner that the "safe" or "off" position is prohibited.
5. Following the application of lockout or tagout devcices to energy-isolating devices, all potentially hazardous stored or residual energy is to be relieved, disconnected, restrained, and otherwise rendered safe. If there is a possibility of reaccumulation of stored energy to a hazardous level, verification of isolation shall continue until the servicing or maintenance is completed, or until the possibility of such accumulation no longer exists.
6. The verification of isolation must be done by an authorized employee prior to starting work on machines, equipment, or pipelines that have been locked out or tagged out.

—Abridged excerpt from M3, *Safety Practices for Water Utilities* (AWWA 2002)

Water utility employees are required to wear personal protective equipment (PPE) in or near the work site. Hard hats prevent possible head injuries caused by falling debris or a backhoe bucket or other moving equipment. Safety gloves protect a worker's hands against cuts or exposure to harmful elements. Safety glasses protect the eyes from being struck by flying objects or discarded material. Hard-toed shoes are vital in preventing damage to the worker's feet.

Parking and storing

All equipment left unattended at night, adjacent to a highway in normal use or adjacent to construction areas where work is in progress, shall have appropriate lights or reflectrors or barricades equipped with appropriate lights or reflectors, to identify the location of the equipment. Whenever the equipment is parked, the parking brake shall be set. Equipment parked on inclines shall have the wheels chocked and the parking brake set.

—Excerpt from M3, *Safety Practices for Water Utilities* (AWWA 2002)

Water providers should hold employees personally responsible for wearing PPE. The equipment reduces or eliminates the injuries that would otherwise occur because of carelessness, inattention, or a lack of proper training in the use of equipment. A safety program that emphasizes the importance of using PPE aids the utility in controlling costs and other negative results associated with loss of time due to injuries.

The safe interaction of equipment, personnel, and materials during the repair process allows for the work to be performed without incident.

Care Around Other Utilities

Damage to underground electric or gas mains can create a dangerous work environment. Severe injury or death could result if proper procedures are not followed. The location and condition of other utilities should be identified during the repair. If work by the water utility damages another utility's equipment, it too must be repaired before the excavation is closed and site restored to its original condition. Sometimes, it may be necessary to turn these repairs over to another utility. Figure 4-5 shows other utilities in the repair excavation.

The other utility must be notified immediately when its infrastructure has been damaged so its personnel can provide instructions for repair or can secure the area until one of its service technicians can be dispatched. The damage and repair of another utility should be documented for reference in the event that an invoice is received and payment requested.

Preparation for Weather Extremes

Extreme weather critically affects the ease with which the leak or break is assessed, the repair process, the backfill effort, and surface restoration. In very cold weather, crews should be mindful of how the temperature will affect these processes. Equipment and materials could be subject to freezing, complicating the repairs and making it more difficult to finish associated tasks.

If workers must spend an extended amount of time in the cold, they should guard against hypothermia. It is helpful to work in shifts and observe one another to make sure no one is overexerting. The wet environment of the leak area, combined with a sweating worker, could cause a cold, the flu, or even pneumonia.

In extreme heat, workers risk heat exhaustion, heat stress, and heat stroke. All three can slow down and even halt the repair process.

Photo courtesy of City of Phoenix Water Services Departmen

Figure 4-5 Excavation of a broken main reveals the proximity of other utility's service lines

Care should be taken to limit or eliminate the negative effect of high temperatures on workers. Workers should stay hydrated and work at a steady pace. A worker who feels light-headed should inform the supervisor, take a break, and drink plenty of water.

During hot weather, quick relief for overheated workers may be found inside the cooler environment of the underground excavation or in an air-conditioned vehicle.

Equipment, fittings, and materials can be damaged during hot weather, although not to the extent they are damaged by cold. Equipment tends to overheat and often shuts down in very hot weather. Metal fittings and pipe, although not themselves damaged, are extremely hot to the touch and may cause burns.

Training employees with an emphasis on safety benefits the water provider in its effort to supply drinking water. Efficiently trained employees tend to operate more safely even when responding during times of emergencies. A safety-conscious supervisor transfers this awareness to subordinate employees. A safety-conscious environment often results in a healthier workforce.

QUESTIONS TO CONSIDER

1. Does the utility have standardized water main repair procedures?
2. If the utility does have repair procedures, how often are they evaluated?
3. Does the utility have an effective training program for repairs?
4. Are the crews equipped to respond to leaks in a timely manner?
5. How often does the utility evaluate the size and configuration of the repair crews?
6. How safely does the utility perform the repair process?
7. Has the organization designated a person responsible for monitoring the safe operation of the work crews?
8. Does the utility have up-to-date safety equipment for all field employees?
9. Does the utility monitor the types of injuries of the employees?
10. Do other utility services often affect the process of repairing leaks?

5

The Process of Repairing Water Main Breaks

The process used to repair a main break should be determined by the extent of the damage, the type of customers affected (e.g., commercial, residential, medical), the needed level of service for the customer type, and other issues, such as quantity of reserve water supplies, traffic impact, and availability of repair and replacement materials.

Prioritizing Repairs

Depending on the size of the workforce and budget limitations, some breaks may need to be prioritized to determine which are most critical and require immediate attention. Priorities can be established in a variety of ways. If water is shooting into the air, that leak clearly needs to be fixed immediately. Certain business communities and commercial districts, such as areas that contain hospitals or hotels, are considered important or high profile, so many utilities respond quickly to breaks in those areas. Repair of other, smaller leaks may have to wait until after hours to lessen the impact on traffic, certain businesses, schools, and local activities.

Many water utilities prioritize their leak repairs based on predetermined service levels or a severity list. To establish a leak severity list to determine which repair should have priority over another, the following questions should be considered:

- Is the main critical to a portion of the system or to the entire system?
- Are any critical institutions on that main?
- Are any special-needs customers in the affected area?
- Did the break result in extensive damage?
- How long has the pipe been leaking?
- What is the leakage rate of the break?
- Are there any major traffic safety concerns?
- Are the necessary repair components readily available?
- Are repair requirements established for service restoration?

Notifying Customers

Prior to the repair, or as part of the process, all affected customers should be notified. The timing of the notification is important. If customers are *not* notified in a timely manner that they will temporarily lack water service, the water provider should be prepared for customer complaints and negative publicity. In rare instances, prior notification may be impractical because of the possible damage to the immediately affected area or to other portions of the distribution system.

If it appears that repairs will be extensive and take several days, affected customers should be advised to store enough water for their essential uses until service is restored. Another approach is to provide customers with an alternate water supply until repairs are completed. Either option will go a long way in an effort to gain and maintain a positive image of the utility.

Customers need to know about the break, the scope of the work to be done, the time frame for repairs, and contact information for the water utility. An employee can inform customers in person by knocking on the door or by leaving a door hanger with these details. Other means of communication include a phone notification system that contacts the homeowner with a message regarding the break, public service announcements on television and radio, and a notification on the utility's website. If the utility has access to its customers' e-mail addresses, a message could be sent this way as well. If the repair work affects a large number of people, a multipronged notification process should be conducted.

The level of communication may vary greatly depending on the extent of the break, the number and types of affected customers, and the impact on other utilities, traffic, and the community at large. An official spokesperson or spokespeople should be designated to communicate a uniform message to the various audiences, including the media. Typically the designated communicator is

- A designated field supervisor,
- A designated departmental representative,
- A public information officer from the utility or the city, or
- Primary field personnel.

Provide an alternate water supply

The water provider should have procedures in place to effectively respond to the water needs of the customer during the repair/restoration process. It is many utilities' policy to provide an alternate source of water if their customers will lack water service for more than a specified amount of time. This may take one of the following forms:

- Free bottled water provided by the agency
- Bottled water purchased by the customer, who is reimbursed by the agency
- Water provided by the agency to a centralized location by water truck or similar means
- The use of bypass piping
- Water provided by an adjacent utility by way of interagency agreement

The spokesperson must be mindful of the audience, particularly customers, the media, regulators, and emergency agencies. Always, the spokesperson must be truthful and avoid speculation. It is better that the correct information—particularly if it is negative—comes from the water utility rather than from the media or consumers.

Emergency response procedures should include a communication component that details the steps necessary to provide critical information at whatever level needed. Breaks that affect an entire community because of the size of the repair or water quality problems may result in a boil-water notice or other communication that must be coordinated with representatives from the regulatory agency and the media.

Notifying Other Utilities

The main requiring attention may be adjacent to another underground utility, such as gas or electric, causing repair or replacement to be more difficult and time consuming. In some instances, a representative from the other utility must be present during the repair or relocation of the other utility's line.

The water utility should be prepared to share the following information with the other utilities:

- How their service will be affected by the break
- How long the service will be affected
- Alternative service options
- A designated contact person for regular updates or changes

Before excavation, marking the locations of all adjacent underground utilities in the public right-of-way (usually performed by a representative of that utility or its contracted agency) is required by law. This helps prevent damage to service lines and interruption of service to other utilities' customers. The process saves money by preventing the loss of other utility services such as electric, gas, or telephone and their repair and restoration. Because the restoration of service by these other damaged utilities is based on the priorities of the impacted utility, it is uncertain if they will be made in a timely manner.

The physical location of the other utilities may affect how much time it takes to makes repairs. When utilities are in close physical proximity to the damaged main or service, the time required for repair generally increases because of the care that must be taken to avoid any interference with other utility services.

Underground utilities that are inaccurately marked or not located at all present unusual problems to the restoration of water service. The unmarked ones, when revealed by their being damaged and/or exposed, will have to be identified by the affected utility and repaired before or after the water repair is made.

Identifying Equipment, Materials, and Personnel

For water service to be restored in a timely manner, it is essential that the needed personnel, repair components, equipment, and support materials are readily available. The size of the repair crew can be determined by its typical functions, the reasonable expectancies for repairs, and the service area covered.

The components required will depend on the extent of needed repairs. In some cases, only a basic repair clamp or the welding of a butt strap may be needed; other situations may require replacement of an entire length of pipe to effectively restore service. The damage to the pipe will dictate what is used for repairs. The following are examples of breaks and leaks:

- A point break (pinhole leak)
- A latitudinal or longitudinal break
- A section or partial blowout as a result of a pressure surge
- Damage to the majority of, or to an entire length of, pipe
- A corrosion leak
- A defective fitting
- A loose bolt or improperly placed rubber gasket

Pinpointing the Location of the Leak

Most leaks are discovered directly beneath the place where water comes to the surface. Some, though, may be some distance away from that point. It is quite expensive and time consuming to excavate in areas with special roadway or walkway pavements, thick asphalt, concrete streets, or even decorative medians without knowing exactly where the leak may be.

The technology of detecting leaks has improved over the years from the early use of the listening stick to ground microphones, leak noise correlators, and more modern approaches. As a result of these improvements, the time spent trying to find leaks has been reduced and accuracy has been greatly improved.

The use of leak detection equipment by staff or a contractor has benefits to the water provider far beyond the potential loss of revenue. The length of time customers are inconvenienced, the discovery of "dry holes," and the fallout of the continued flow of water until the leak is actually found creates or exacerbates a lack of confidence from the community being served.

Excavating for Leak Exposure

The value of a good equipment operator is revealed during the excavation process. Skillful operators are sensitive enough to underground utilities and obstructions that they are able to expose the damaged main or service effectively. Less competent operators can cause additional problems by damaging another utility's infrastructure or further destroying the water line being exposed. In the latter case, the job takes longer and customers lack service longer.

It is sometimes beneficial to maintain reduced pressure on the line until the leak has been discovered. Certain breaks or leaks may be on the bottom of the pipe and so are difficult to identify unless positive pressure ensures continuous flow. If this process causes safety concerns, a further reduction in pressure should be considered. In the case of larger-diameter pipe and in the interest of safety, the pressure should be removed and/or eliminated during the repair. Whenever the flow of water has been stopped completely or when there is a possiblity of contaminants entering the pipeline, the pipe must be chlorinated before it is placed back in service.

Where practical and to reduce the possible introduction of a contaminant, the area around the pipe should be excavated to a level beneath the pipe. This keeps debris and water from the break at a level that prevents it from flowing back into the main.

Sanitary repair practices

Care should be taken to prevent water, dirt, and other material from entering the damaged pipeline. Surface water should be diverted from the construction site using barriers if necessary. The excavation should be dewatered so that water is below the pipe invert. Before beginning a repair, the interior of the pipe should be cleaned where it has contacted the soil or backfill material. Soil beyond reach can be removed by low velocity flushing in both directions. If possible, the flow or positive pressure should be maintained in the leaking pipeline to prevent backflow of contaminated water. This is particularly important until the leak is exposed and secured.

Repair crews need to be aware that their actions could contaminate the water supply during a pipeline repair. Crews must ensure the cleanliness of all cables, pipes, and hoses drawn through the inside of the pipelines or appurtenances. Disinfection (using chlorine dip or swab solutions) of hand tools (such as saws) used in the repair is a prudent step. Equipment operators should also use caution to avoid introducing soil or other material into the repaired pipeline.

—Excerpt from *Disinfection of Pipelines and Storage Facilities Field Guide* (Lauer 2006)

Evaluating the Pipe Damage and Leak

It is important to inspect not only the pipe but the soil around it and any adjacent connections, fittings, and/or structures when evaluating the cause of the leak and condition of the pipe. The soil around the pipe or fittings may be corrosive and instrumental in the process of degradation. This is often seen in various areas and can be validated as a result of many utilities performing soil resistivity examinations as well as requiring the placement of corrosion inhibitors/resistant coatings and processes as part of the pipe repair/installation procedures. Some of the other causes of leaks were covered previously in this guidebook.

The examination of the main and adjacent area can also be used as a teaching opportunity for trainees and/or newer employees. The kinds of breaks and their causes can be discussed, as well as some inhibitive or preventive measures. All aspects of job-site and trench safety should be observed.

Multiple or Complex Leaks

When the leak is exposed and several previous repairs are evidenced by repair clamps or other repair devices a short distance from the latest break (see Figure 1-1), it is vitally important to consider replacing a larger portion of the pipe or even the entire section depending on its condition.

The costs of several repairs to the main versus the cost to replace the damaged or substandard section of pipe should be considered. Every time a crew is dispatched to the location, customers' service is disrupted. After several times, the customers tend to become more agitated about the process and begin to complain.

Based on the service restoration processes, the availability of the crews, the total cost of repairs, and the utility's concern for the inconvenience of customers, a utility must decide when it is more cost-effective to replace the main or service instead of continuing to make repairs.

The numerical portion of this evaluation is as simple as determining the entire cost to replace the damaged section(s) of pipe and comparing it against the sum of all of the individual repairs made over a specific period of time. Many water providers have a number of breaks per certain length of pipe to use as part of their decision making process.

The simple question is, "How many times is the pipe repaired before the replacement is justified?"

Complex Main Breaks

A similar concept holds true when evaluating complex main breaks, where the repair will result in greater impacts to other portions of the main, customers, or adjacent structures. These decisions must be viewed not only in light of existing conditions but also along with any planned system expansions.

Because they often affect a greater portion of the system than a typical main break, complex breaks require more planning, preparation, and actual physical work than basic repairs and replacement. The following lists other types of complex breaks that should be considered and planned for before they happen.

- A break in which more than one section of pipe/fitting is involved—for example, a longitudinal crack, an offset joint, a cross, or a transition piece
- A break at a fire hydrant branch line or service connection
- A break on a large-diameter pipe at a tapping-valve connection
- A break adjacent to, just above, or just below another utility

Figure 5-1 shows a break adjacent to a cross fitting. The repair required more effort and materials than would usually be the case because of the existence of other utility lines in the immediate area.

Photo courtesy of City of Phoenix Water Services Department

Figure 5-1 A main break at a fitting

Dissimilar Pipe Materials

The variation of pipe material creates a slightly different repair response in many situations. If there is a leak on a section of small-diameter (4 in.–12 in.) asbestos–cement (AC) pipe, for example, the pipe is often replaced with ductile-iron pipe (DIP) or polyvinyl chloride (PVC) using transition couplings. (Note: State and federal regulatory agencies have requirements on how to handle AC pipe, including burying it in place with a barrier to identify it and placing it in a plastic bag or burlap sack before removing it from the hole.) Cast-iron pipe replacements are made using DIP or PVC as well. Galvanized and/or lead service lines are replaced by copper tubing or plastic pipe.

Making repairs on pipe 16 inches in diameter and larger requires a different approach than repairs to pipe that is 12 inches and smaller. The type of pipe material affects the overall approach as well, particularly for the larger sizes. Consider large-diameter ductile-iron pipe compared to a large diameter PVC pipe. The size and weight alone could result in different equipment being used, additional personnel, and other repair components. For example, when sections of large-diameter concrete pipe need repair, replacement pieces must be moved into place using equipment that is larger than the typical backhoe that is used for smaller diameters and lighter weight pipe. Heavier pipe is usually brought into the site on a flatbed truck and has to be offloaded and placed onsite. This process may require a large area for storage and staging. Figure 5-2 shows a large pipe that required heavy duty equipment to be exacavated and moved.

On steel pipe, a point repair of pinhole leak usually calls for a patch to be welded over the hole and a coating applied. When a larger section needs repair or replacement, butt straps are used to secure the new section(s) to the existing pipe. HDPE pipe repairs require a specialized process known by the installers and those responsible for repairs.

Consideration should be given to the repair or replacement process when connecting dissimilar pipe materials. Different metal pipe materials that are connected as a result of a repair may need corrosion control mechanisms to prevent degradation or reduce its rate. Wherever practical, it is better to totally replace the defective material as part of the process.

Photo courtesy of Denver Water

Figure 5-2 A damaged large-diameter pipe removed from the excavation.

Fittings

A repair component that will be less than two to three pipe diameters (depending on the size of the main) to another fitting, such as a 90-degree elbow, a cross, a tee, a valve, a service, or a pipe joint, should be evaluated and an alternate repair process should be considered. The closeness of the fitting to the repair can result in increased pressure and air at the fitting as water moves through the pipe and exits at the fitting. One option is to replace the section of pipe from the fitting to a distance farther away or to the next joint. Relocating the fitting should be considered as well. If thrust blocks are present, caution should be taken not to disturb the soil around them; if a thrust block's stability is compromised, it should be replaced. Figure 5-3 illustrates the importance of using a thrust block behind a connection to another pipe.

What is a high-risk pipe?

A "high-risk pipe" is a pipeline constructed using dated materials or methods, often made of cast iron, running through an area with heavily trafficked street crossings. This is a perfect description of a typical trunk main in many urban centers. These urban centers typically represent significant loss potential from damage caused by water main breaks. This damage occurs because of a high density of buildings; concentrated underground infrastructure (e.g., utilities, transit systems, parking garages); important traffic thoroughfares; and economic loss potential from outages of power, gas, and other utilities damaged during rupture events. Many pipelines at significant risk for failure exist where the consequences of failure are most dramatic in terms of liability to the water utility.

Risk factors for pipe breaks include the following:

- cast-iron pipe material (used heavily in older high-pressure pipes);
- lead-sulfur–based joint compounds (leadite), commonly used from the 1930s to the 1950s;
- heavy traffic load over the pipeline;
- pressure transients;
- pipe age, which is associated with outdated materials and installation practices; and
- impacts from construction in the vicinity.

Some failure mechanisms operate over very long periods of time. The result—older pipes experiencing these conditions are more likely to fail. However, there are other factors at work: material or installation defects may surface over a relatively short period of time; problematic regions of a pipeline may be replaced or rehabilitated; or third-party damage can cause a previously healthy pipe to degrade. The net result is that age alone cannot be relied on as an indicator of an at-risk pipe.

—Excerpt from Kunkel et al., 2008

Chlorinating

If the main is damaged to the point that excavation may introduce contamination to the water, or if the pressure was reduced below the minimum required (typically 20 psi), contamination may have been introduced. Therefore it is important to chlorinate as required by the appropriate regulatory agency.

One approach is to chlorinate the immediate and adjacent areas of the break. This can be done by using a pressurized spray container with a chlorine solution within the acceptable range required by regulation. Another approach is to chlorinate the entire section of pipe from valve to valve or valve to dead end. The amount of chlorine needed to disinfect the main depends on the length of pipe exposed or replaced, or the section of pipe that was isolated during the repair.

Photo courtesy of City of Phoenix Water Services Department

Figure 5-3 The lack of a thrust block at the pipe bend contributed to a blowout that caused significant road damage

The main needs to be recharged so normal operating pressure is restored, while effectively preventing or managing any surges that could cause another leak. To aid in this process, an adjacent fire hydrant should be opened to release any air in the line. In some instances, one or more service lines can be used. As a general rule, mains should be recharged by moving air in the line from a lower elevation to a higher one so the air can be discharged into the atmosphere through an air release valve, a fire hydrant, or a service line.

Another critical part of restoring a main to service is dechlorinating the main. Removing chlorine from the water before it is released to a storm sewer is an accepted or required practice for most utilities to protect aquatic life downstream. Some water reclamation districts have specific requirements for removing chlorine before they allow the water to enter their systems. In some instances, if the water can be captured in a sanitary sewer main, dechlorination may not be required. In any case, coordination with the wastewater utility should be part of the process to mitigate any negative impact to its wastewater treatment plant.

AWWA has numerous books, manuals, standards, and videos that cover the chlorination process in detail. Most state or provincial environmental or water quality departments may have or make reference to specific procedures as part of their requirements. A good resource that describes the proper procedures for returning a pipe to service is *Disinfection of Pipelines and Storage Facilities Field Guide* (Lauer and Sanchez 2006).

Providing and Compacting Proper Backfill Material

The proper placement of select backfill material reduces the amount of settlement experienced by the soil surrounding the pipe and adjacent components. As part of this process, municipal utilities have specific requirements for compaction that are intended to eliminate settlement.

Some utilities require the use of flowable fill to expedite the restoration process, in high traffic areas, where interfering underground utilities are located, and in some cases as a typical response. This material provides the stability and strength necessary to protect underground utilities, yet it is pliable enough to be excavated without much difficulty. It enables faster surface restoration so that traffic flow can be restored sooner. Figure 5-4 illustrates a front-end loader placing flowable fill into an excavation.

Water pipe laid in areas prone to ground movement needs to be stabilized along the pipe, joints, and connections. The repair is often costly and usually requires mitigation over a large surface. Figure 5-5 illustrates this kind of leak. Notice that the asphalt has washed out along with the fill material and left a void in the area around the pipe. If the water had not surfaced or the void was not discovered and the hole was left unattended, a vehicle could have driven into it, causing damage and possible injury to passengers.

To reduce or eliminate problems associated with settling soils, the new pipe should be positioned properly, and sufficient support and backfill beneath, around, and above it should be provided. Poor

Photo: Bryan Bechtold, AWWA

Figure 5-4 Front-end loader fills a an excavation with flowable fill.

Figure courtesy of City of Phoenix Water Services Department

Figure 5-5 Washed-out fill as a result of a water main break

compaction of the fill dirt could result in soil movement and ultimately another break or leak.

Application of the surface treatment concludes the repair. Below certain temperatures and at certain times of the year, asphalt batch plants are closed because asphalt normally does not place well as a permanent patch at or below freezing temperatures and hot patches for surface restoration are unavailable. Some municipal utilities use steel plates to cover holes until they can restore the surface properly.

Performing Proper Site Restoration

Restoring the site to its previous condition is as important as the break repair itself. The attention and care given to this part of the response reveals the level of concern for the customer. Paved areas, curbs and gutters, and sidewalk sections should be restored. Grass, shrubs, and trees should be replaced, as should fences, block walls, and other utility aboveground or underground components.

When performing this function on customers' property or in the right-of-way adjacent to their property, it is important to reach an agreement with them about what gets placed where prior to the work. For example, if a tree or shrub needs to be removed to access the pipe for repair work and possibly for future access, there should be a clear understanding of where the foliage is to be replaced. This reduces the likelihood of any misunderstanding once the work is completed. If a landscaping company is hired to restore the site, it should also follow these guidelines.

QUESTIONS TO CONSIDER

1. How does the utility select and purchase repair equipment?
2. Is there an ample supply of repair/replacement parts at the utility?
3. Does the utility evaluate the materials and equipment used in the repair process?
4. How knowledgeable are the employees of the procedures for repairing main breaks?
5. How are the breaks/leaks prioritized for repair?
6. Does the utility have any requirements for the closeness of fittings and/or connections?
7. Does the utility have requirements for the placement of water mains?
8. Is an evaluation made of the pipe adjacent to the break/leak?
9. When does the utility consider replacement over repair?
10. Does the utility have any repair/replacement agreements with other utilities?
11. How quickly are other utilities located when requested?
12. If repairs are made by a contractor, what are the components of the agreement?
13. Does the utility monitor the repair work performed by contractors?
14. Does the utility have site mitigation procedures?

6

Documenting Water Main Breaks

During a repair, water professionals are concerned with restoring service to customers and getting things back to normal. This is also a great time to gather important information about the pipe itself, system components that are vital to isolating the break, the soil adjacent to the pipe, and staff, equipment, and material readiness. This effort contributes to an accurate accounting and mapping of the quantity, kinds, and locations of pipe materials in a utility's distribution system and should become part of the general documentation about the water system and its condition.

Documenting repairs and evaluating processes provide information that aids in planning for current and future events; help determine whether to replace a pipe, repair it, or rehabilitate it; and are invaluable for justifying capital investments.

When the pipe is exposed so the cause of the break can be assessed, the following information should be collected:

- The exact location of the pipe
- The type of pipe material
- The overall condition of the pipe and its adjacent appurtenances

- Any previous repairs
- The condition of the surrounding soil
- The location of other utility service lines
- Water quality parameters

Documenting the conditions and the work processes also provides utility managers a way to confirm the effectiveness of the repair staff and provides repair crew members with valuable information to improve their response. Otherwise, they may continue to use the wrong components, and the real problem may not be determined. In addition, documenting all construction activity when installing water pipe provides a legal record if a problem arises regarding a contractor's poor practices.

Identifying and Compiling Critical Information

To best identify what information to collect, the water utility should first determine how it will use the collected data. For example, both rehabilitation and replacement programs require the use of main break information to justify their existence. Some utilities use an established amount, budgeted annually, for these programs and set priorities depending on the amount required for a given project and the budget constraints. Others incorporate their repair budget into the overall operations and maintenance budget without projecting specifically how the money will be spent.

Data about water main breaks should become part of the water utility's historical record of its assets. This information can also help with determining a current valuation of the distribution system; data assessing the current condition of the system provides a more realistic evaluation of the pipe than its stated design life.

A description of the pipe material and size should be added to the database, as well as data about the condition of adjacent pipe and appurtenances. Geographic information system/global positioning system (GIS/GPS) technology should be used to accomplish this. Knowing the coordinates of valves, mains, fire hydrants, and other critical components has great value if an intersection or area is flooded and all access points are underwater and not visible. GIS/GPS locators can greatly reduce the amount of time necessary to close or open important valves during the emergency response.

Relatively inexpensive GIS/GPS equipment is available that can be used during scheduled O&M to identify components' coordinates. Valve coordinates can be obtained during the valve assessment/evaluation process and stored in a database. While performing maintenance on a fire hydrant, its coordinates can be acquired and added to a database as well.

Another advantage of being able to quickly identify the number and locations of fittings and any associated problems is apparent when discussing problems about a product with a manufacturer's representative. Often, several water providers experience the same difficulty with a product, revealing a problem that should be corrected throughout the industry.

Asset Management

Repair, rehabilitation, and replacement options must be part of a utility's toolbox for maintaining adequate service and supply. When properly evaluated, main break data can help to identify whether repair or replacement is the better approach. This is basic to asset management. To determine the better option, identify the total average cost of a single repair for the various sizes of pipe (see sidebar). A benefit–cost ratio can be calculated to determine the more appropriate option.

Next, determine how many single repairs can be made within close proximity before it becomes financially more practical to replace the entire section of pipe. This part of the process is important because often, and less cost-effectively, utilities just make individual repairs as needed without including repairs that have recently been made in the same area or that they are close together.

If it is determined, for example, that three repairs within a 100-foot section of the pipe is the maximum before the replacement option becomes more cost-effective, then before or during the second repair, preliminary preparations should be made for pipe replacement. Such tasks would include identifying how much pipe must be replaced, who will do the work, and when. Pipes and parts may need to be ordered. In addition, customers will have to be notified. Alternative water supplies or bypass piping will have to be arranged.

Much can be said about trying to reduce the number of times customers' service is interrupted. To customers, a service disruption is just that, regardless of the reason. Patience begins to wear thin when people are inconvenienced by repairs that should have been made that would have prevented the current disruption. If the utility chooses to replace the pipe, continuing customer inconvenience can be avoided and customer satisfaction increases.

It is in the best interest of the water provider and ultimately the customer if the utility uses the information gathered to justify rehabilitation/repair/replacement programs. Implementation of these kinds of processes should be proactive and focused on adding value to the failing infrastructure. Annual funding should be established for each approach to provide a framework to maintain and enhance existing components. Every water provider should establish a goal to annually replace a certain percentage of its system that has reached its useful life.

Repair, rehab, replace?

There is no one-size-fits-all approach to replacement and repair. As long as each approach is appropriate from a cost–benefit standpoint, it can work effectively. Take these issues into consideration when determining whether to rehab or to replace:

- The break/leak history of the pipe
- The size of the existing main
- The location of the existing pipe—heavy traffic areas, rights-of-way, and so on
- The aesthetic condition of the surface above the pipe—historic pavement, tree lined medians, and so on
- Design versus actual flows and pressures
- The condition of critical water quality parameters—turbidity, chlorine residual, and so on

Leak Reports

On the leak report form, a utility can capture and document critical information. Just as with repair data, the way it plans to use the information collected during a leak report and investigation determines what is documented.

Some utilities store the data just as a matter of record. Others use it more effectively to help shape a repair and replacement schedule to enhance and improve their infrastructure. Leakage can have a serious impact on a utility's bottom line, and by reducing it savings can be found in power generation, maintenance, and treatment costs. When the information helps to determine where capital improvement funds are spent, which rehabilitation option is considered, or when to replace the pipe instead of making a point repair, it is more valuable than just a historic record.

AWWA Manual M28, *Rehabilitation of Water Mains*, covers this topic in great detail and provides a list of rehabilitation options based on the listed items and other critical considerations. The rehabilitation options provided are nonstructural, semistructural, or structural; they can be determined by the evaluation process covered in that manual.

Analyzing and comparing the costs to repair versus replace the pipe yields the best option available. To accomplish this properly, all costs associated with each option must be part of the procedure. This includes costs of all indirect or supportive activity to totally complete the work.

One of the best reasons to replace the damaged main is if it is undersized. This provides the ability to increase the pipe size to one that better serves the existing customers, enhances the agency's approach to move more water into and out of selected areas, and can often improve pressure. Finally, a larger main helps the water utility to provide more water for fire protection.

Many water utilities are required to request rate increases before boards or commissions. Having solid data about water main leaks or breaks helps officials to present an accurate and compelling case based on the condition of the infrastructure documented by a report that identifies the causes of leaks and breaks along with their frequency.

AWWA's manual M36, *Water Audits and Loss Control Programs*, provides great detail about the importance of controlling water loss through leakage and how to conduct water audits that identify where and how water is being lost.

Responding to Regulators

To confirm that drinking water is of high quality after water main repairs have been made, regulatory agencies may require the following information:

- The type of break
- The impact on customers
- The possibility of exposure to contamination
- The extent of the water outage
- The results of sampling after the repair/replacement
- The use of notices and notifications

The regulatory agencies require that documents be maintained for a certain number of years based on the reporting structure established, the importance of the information to the operation, the type of system, and so on. Information about main breaks aids the utility to better manage its assets and demonstrates to regulators a professional level of responsibility.

If the utility is audited by regulators, the possession of a main break database containing all of the information previously identified reflects its commitment to success. The database can also be used to develop a rehabilitation or replacement program.

Finally, collecting data about the conditions at the scene of water main break can be useful to the water provider when claims are filed against it. Noting the condition of the street, sidewalk, or driveway can reduce or eliminate paying for damages or the extent of damage not caused by a main break.

QUESTIONS TO CONSIDER

1. Is the utility equipped to receive and act on information from customers about main breaks?
2. Does the utility provide and receive comment cards from the customers' evaluations of the repair process?
3. How does the utility document the information it receives?
4. Does the utility document soil conditions?
5. Does the utility have programs directed at monitoring, evaluating, and reducing leaks?
6. Does the utility use customer concerns to improve the leak response program?
7. Does the utility use a leak report document as a water main evaluation tool?
8. How is the leak information database accessed?
9. Who is responsible for maintaining the information?
10. How does the utility budget for rehab/repair/replacement programs?
11. What is the amount of water saved as a result of the utility's leak detection/audit programs?
12. Does the utility evaluate the operational impact of main breaks on the system?
13. What are the components of the utility's asset management program?

Appendix A

Pipe Repair Checklist

As published in July 2008 *Opflow* Question of the Month

Thanks to: Mike Grahek, Los Angeles Department of Water, Los Angeles, Calif.; Ahmad Havivian, Black & Veatch, Gaithersburg, Md.; David Kroon, Corrpro, Houston, Texas; David Lippman, Las Virgenes Metropolitan Water District, Calabasas, Calif.; Ken Morgan, Water Distribution Superintendant, Phoenix, Ariz.

Pipe Repair: Before The Work Commences

- Employee training:
  - Confined spaces
  - Lockout/tagout of valves and hydrants
  - Traffic awareness
  - Proper excavation, trench work, and backfilling techniques
  - Handling and installing pipe material
  - Hand signals
  - Chemical safety and handling
  - First aid and cardiopulmonary resuscitation (CPR)
  - Proper sampling, chain of custody, and transport techniques for bacteriological tests

- Additional training for project managers and supervisors:
  - Requirements for disinfection, dechlorination, and disposal after repairs that may vary by locality
  - Regulations pertaining to hydrostatic testing and bacteriological sampling
  - Required documentation and records retention
- Distribute advance notification of repair (including scheduled periods of work, expected traffic disruptions or service interruptions, and expected public hazards) to:
  - Department of Public Works
  - Department of Transportation
  - Local law enforcement
  - Local government leader(s)
  - State Department of Health (if applicable)
  - Media (print, radio, television, websites), in the form of a press release
  - Affected customers

Pipe Repair: After The Work Ends

- GPS or other notation of the exact location of underground water mains, services, valves, hydrants, and other underground utility locations should be noted clearly on a system map.
- Notation of OD, pipe material, manufacturer, date of manufacture, and supplier information should be included in the project file.
- Determine (or estimate) whether repair is due to leak, break, beam, split, blowout, joint, corrosion, or other damage. If possible, estimate the date the pipe was installed, and the cost associated with the repair (materials, manpower, time, overtime, etc.) for use in future budgeting activities.
- Fill out the Main Break Evaluation form.

Pipe Repair Checklist

- Distribute notifications of repair work.
- Place public warning and road hazard signs, traffic cones, and barriers.
- Include night work lights (if applicable).
- Check personnel protective equipment (gloves, reflective vests, hard hats, etc.)
- Provide surface runoff diversionary equipment (sandbags, trench covers, and the like).
- Provide dewatering pump and accoutrements.
- Bring flow diversion equipment.
- Include disinfection/dechlorination chemicals and equipment.
- Bring pressure measurement gauges or devices.
- Provide tapping equipment.
- Include hand tools (saws, wrenches, buckets, shovels, pick axes, ladders, flashlights, and the like).
- Provide biological sampling bottles, protective gloves, transport cooler, ice packs, and laboratory chain-of-custody sheets.
- Notify appropriate parties of repair (in advance if possible).
- Determine how groundwater levels, inclement weather, and other factors may affect the repair, and determine compensatory methods.
- Call "dig safe" system (if available) to identify buried utilities in the area.
- Locate, mark, and protect existing underground utility lines (including water, sewer, phone, cable, gas, and power lines) and valves in the vicinity of the repair.
- Determine location of dewatering and runoff, and avoid or mitigate erosion and property damage.
- Determine whether temporary service can be provided (if applicable).

- Isolate the pipe section with valves, maintaining positive pressure to reduce backflow or runoff contamination.
- Expose, thoroughly scrape, and clean the area around the pipe section. Inspect breaks for rough or sharp edges and file smooth.
- Determine the pipe material and its outside diameter (OD), fittings, joints, gaskets, clamps, and other repair equipment necessary to perform work. If work takes more than one day, store materials on-site in a secure area.
- Disinfect hand tools and tapping machines used in the repair prior to use; clean and disinfect parts and materials before installation. Maintain protective caps, plugs and coverings on pipes and fittings until just prior to installation, to prevent contamination.
- Perform repair.
- Flush, clean, and disinfect the repaired pipe.
- Complete repair; tighten and/or secure installation.
- Apply corrosion protection (if necessary).
- Perform hydrostatic testing per applicable AWWA pipe standard, and bacteriological testing as according to AWWA C651: *Disinfection of Water Mains.* Obtain appropriate permission to return line to service.
- Dispose of or dechlorinate chlorinated water according to regulations.
- Flush hydrants to remove debris.
- Sequence valve operations so that negative pressure in the system is avoided.
- Backfill and compact pipe bedding per applicable AWWA pipe installation standard; repair ground surface to at least original condition.

FIELD DATA FOR MAIN BREAK EVALUATION: PROJECT NUMBER ______

Date of Break: ____________________ Time: ________ ☐ A.M. ☐ P.M.

Type of Main: ____

Size (OD): __________ in. Joint: ____________ Cover: ____ ft ____ in.

Wall Thickness at Failure: __________ in.

Nature of Break:

☐ Circumferential ☐ Longitudinal ☐ Both

☐ Blowout ☐ Joint ☐ Sleeve ☐ Split at Corporation ☐ Other

Apparent Cause of Break:

☐ Water Hammer (Surge) ☐ Defective Pipe ☐ Deterioration

☐ Corrosion ☐ Improper Bedding ☐ Operating Pressure

☐ Temperature Change ☐ Differential Settlement ☐ Contractor

☐ Other: ____

Pipe Location Information:

☐ Paved ☐ Unpaved Traffic: ☐ Heavy ☐ Medium ☐ Light

Type of Street Surface: ______________________ Side of Street: ☐ Sunny ☐ Shady

Type of Soil: ______________________________ Resistivity: ________________ ohms/cm

Electrolysis? ☐ Yes ☐ No Corrosion? ☐ Inside ☐ Outside

Other: ☐ Rocks ☐ Voids Proximity to Other Utilities: ________________

Depth of Frost: ________ in. Depth of Snow: ________ in.

Comments: ____

____

____

OFFICE DATA FOR MAIN BREAK EVALUATION: PROJECT NUMBER ______

Weather Conditions During Previous Two Weeks: ______________________________

Sudden Change in Air Temp: ☐ Yes ☐ No Average Temp: ____ °F Rise: ____ °F Fall: ____ °F

Sudden Change in Water Temp: ☐ Yes ☐ No Average Temp: ____ °F Rise: ____ °F Fall: ____ °F

Historical Pipe Data:

Type of Main: __________ Class/Thickness: __________ Laying Length: __________ ft

Date Laid: __________ Design Operating Pressure: ____________________ psi

Previous Break Reported? ☐ Yes ☐ No Pressure at Previous Break? ________ psi

Historical Installation Data:

Bedding: ☐ Gravel ☐ Sand ☐ Native Material ____________________

Backfill: ☐ Gravel ☐ Sand ☐ Bank Run Sand and Gravel ☐ Rock

☐ Native Material: ______________________________

☐ Other: ______________________________

Compaction: ☐ Natural ☐ Water ☐ Compactors ☐ Vibrators

☐ Other: ______________________________

Comments: ______________________________

____

____

____

OTHER DATA

Location of Break: ______________________________ Map Number: ______

Reported By: ______________________________

Damage: ______________________________

Repairs (Equipment, Materials, Labor): ______________________________

Repair Difficulties (if any): ______________________________

Installing Contractor: ______________________________

References And Bibliography

American Society of Civil Engineers (ASCE). 2009. Report Card of America's Infrastructure. Reston, Va.: ASCE.

AWWA (American Water Works Association). 2012. *Buried No Longer, Confronting America's Infrastructure Challenge.* Water Industry Technical Action Fund Report. www.awwa.org/files/GovtPublic Affairs/GADocuments/BuriedNoLongerCompleteFinal.pdf

AWWA. 2011. Manual of Water Supply Practices M58, *Internal Corrosion Control in Water Distribution Systems*. Denver: AWWA.

AWWA. 2010. Principles and Practices of Water Supply Operations, *Water Transmission and Distribution*, 4th ed. Denver: AWWA.

AWWA. 2010. ANSI/AWWA Standard G200, Distribution Systems Operation and Management. Denver: AWWA.

AWWA. 2009. Manual of Water Supply Practices M36, *Water Audits and Loss Control Programs*, 3rd ed. Denver: AWWA.

AWWA. 2005. *Water Distribution Operator Training Handbook*, 3rd ed. Denver, Colo.: AWWA.

AWWA. 2005. ANSI/AWWA Standard C651, Disinfection of Water Mains. Denver: AWWA.

AWWA. 2004. Manual of Water Supply Practices M27, *External Corrosion: Introduction to Chemistry and Control*, 2nd ed. Denver: AWWA.

AWWA. 2002. Manual of Water Supply Practices M3, *Safety Practices for Water Utilities*, 6th ed. Denver: AWWA.

AWWA. 2001. Manual of Water Supply Practices M19, *Emergency Planning for Water Utilities*, 4th ed. Denver: AWWA.

AWWA. 2001. Manual of Water Supply Practices M28, *Rehabilitation of Water Mains*, 2nd ed. Denver: AWWA.

Awwa Research Foundation. 2001. *Good Practices to Prevent Microbiological Contamination of Water Mains and Field Pocket Guide.* AwwaRF Project 2610. Denver, Colo.: Water Research Foundation.

Bloetscher, F. 2011. *Utility Mangement for Water and Wastewater Operators.* Denver: AWWA.

CDC. (Centers for Disease Control and Prevention) Division of Oral Health, National Center for Chronic Disease Prevention and Health Promotion. 2009. Fluoridation of Drinking Water and Corrosion of Pipes in Distribution Systems. www.cdc.gov/fluoridation/fact_sheets/engineering/corrosion.htm.

Eisenbeis, P., P. LeGauffge, and S. Saegrov. 2002. Water Infrastructure Mangement: An Overview of European Models and Databases, in *Assessing the Future: Water Utility Infrastructure Management,* D.M. Hughes, editor. Denver: AWWA.

Grigg, N.S. 2004. Assessment and Renewal of Water Distribution Systems. AwwaRF Project 2772. Denver, Colo.: Water Research Foundation and AWWA.

Kerri, K. 2010. *Water Distribution System Operation and Maintenance–A Field Study Training Program*, 5th ed., Sacramento, Calif.: California State University–Sacramento, Office of Water Programs.

Kunkel, G., K. Laven and B Mergelas. 2008. Does Your City Have High-risk Pipes? *Journal - American Water Works Association.* 100(4):70–77.

Lauer, W.C. and F.J. Sanchez. 2006. *Disinfection of Pipelines and Storage Facilities Field Guide.* Denver: AWWA.

NACE International. 2007. Standard SP0169-2007, Control of External Corrosion on Underground or Submerged Metallic Piping Systems. Houston, Texas: NACE International.

Nance, S. MWRA clears boil-order for 2 million in 59 hours. *Streamlines*, May 11, 2010. www.awwa.org/publications/Streamlines Article.cfm?itemnumber=54428

Parmelee, M.A. 2004. Domino theory accounts for Boston main failure. *Streamlines.* May 31, 2011. www.awwa.org/publications/Stream linesArticle.cfm?itemnumber=56856

Romer, A.E., and B. Bosserman II. 2009. *Corrosion Control for Buried Water Mains Pocket Field Guide.* Denver: AWWA.

US Department of Transportation. 2009. *Manual on Uniform Traffic Control Devices for Streets and Highways.* Federal Highway Administration. mutcd.fhwa.dot.gov/pdfs/2009/pdf_index.htm

US Environmental Protection Agency (USEPA). 2007. Innovation and Research for Water Infrastructure for the 21st Century Research Plan. www.epa.gov/nrmrl/pubs/600x09003/600x09003.pdf

USEPA. 2003. Summary of the European Experience Using Non-Hydraulic Models forInfrastructure Rehabilitation. Decision-Support Tools for Predicting the Performance of Water Distribution and Wastewater Collection Systems.

USEPA. 2002. The Clean Water and Drinking Water Infrastructure Gap Analysis Report. www.epa.gov/awi/pubs/816r02020/816r02020.pdf

Index

About the Author

Kenneth C. Morgan, PE, has been actively involved in the water industry for more than twenty-five years. He has extensive knowledge in water distribution system component evaluations, operator certification training, and water security. He has advised several municipal agencies and water providers in operation and maintenance program development and implementation.

Ken has a bachelor of science degree in civil engineering from the Missouri School of Science and Technology (formerly the University of Missouri–Rolla). He is a registered professional engineer in Arizona, Colorado, and Missouri. Ken holds a Grade IV Water Distribution System Operator certificate in Arizona.

The author can be reached at

KCM Consulting Services LLC
P.O. Box 10419
Phoenix, AZ 85064
e-mail: kenmorgan@h2otrainer.com
Phone: (602) 708-8412

www.ingramcontent.com/pod-product-compliance
Lightning Source LLC
LaVergne TN
LVHW022112200926
848481LV00003B/1158
* 9 7 8 1 5 8 3 2 1 8 6 0 0 *